Preface

There are many good books. But it's difficult to get interesting teaching materials simply written. I'm planning a new series of school textbooks. It has the following features:

- written in a clear and concise
- The example contains concept and application
- Bilingual in English and Chinese

Do each question yourself   This is the best way to learn

I hope this book would give much help to the readers

Shi Kwok Wong

序

　　好書很多　然而耍尋得簡明有趣的教材也決非易事
我已考慮寫一套新的系列教學參考書.它應具有下述
特色:

- 簡明易懂
- 例題包涵概念和運用
- 英中雙語參考

　　動手做每一題　　　這才是最好的學習方法

　　我希望本書對讀者能有所幫助

　　　　　　　　　　　　王士國

Contents

目錄

1 Positive integer

Definition

The following numbers

1,2,3,4,5,6,7,8,9,10

11,12,13,14,15,16,17,18,19,20

....

101,102,103,104,105,106,107,108,109,110

111,112,113,114,115,116,117,118,119,120

....

 are called Positive integers

And each following number

1,2,3,5,7,11,13,17,19,........101,103,107,109,113

....is a prime

And each following number

4,6,8,9,10,12,14,15,16,18,20,........118,119...........

is a double prime

So   positive integer contains prime   double prime

Note

we consider it for the following reasons

2 Factors

Definition

Multiplier is the factor of product

Example 2-1 $\quad 120 = 8 \times 15 = 2^3 \times 3 \times 5$

So $\quad$ 8, 15, 2, 3, 5 is a factor

Sometimes a number contains many factors $\quad$ It's no easily to find all factor

In fact

$120 = 2 \times 2^2 \times 3 \times 5 = 2 \times 4 \times 3 \times 5$

$C_4^1 + C_4^2 + C_4^3 + C_4^4 = 4 + 6 + 4 + 1 = 15$

$C_4^1 = 4 \rightarrow 2, 4, 3, 5$

$C_4^2 = 6 \rightarrow 2 \times 4, 2 \times 3, 2 \times 5, 4 \times 3, 4 \times 5, 3 \times 5$

$\qquad \rightarrow \quad$ 8, $\quad$ 6, $\quad$ 10, $\quad$,12 $\quad$,20, $\quad$ 15

$\qquad\qquad C_4^3 = 4 \rightarrow 2 \times 4 \times 3, 4 \times 3 \times 5, 3 \times 5 \times 2, 5 \times 2 \times 4$

$\qquad \rightarrow \quad$ 24, $\qquad$ 60, $\qquad$ 30, $\qquad$ 40

$C_4^4 = 1 \rightarrow 120$

So $\quad$ 2,4,3,5,8,6,10,12,20,15,24,60,30,40,120

ia a factor

Remember

A number contains many factors $\quad$ it's no easily to find each factor

3   Prime

Definition

A positive integer which is divisible only by 1 and itself is called prime

A positive integer which is divisible by 1, itself and other factor is called double prime

Example 3-1 $5 = 1 \times 5$ $5 \div 1 = 5$ $5 \div 5 = 1$

5 is a prime

Example 3-2 $113 = 1 \times 113$

$113 \div 1 = 113$

$113 \div 113 = 1$

113 is a prime

Example 3-3 $10 = 2 \times 5$

$10 \div 1 = 10$

$10 \div 10 = 1$

$10 \div 2 = 5$

$10 \div 5 = 2$

10 is a double prime

Example 3-4 $114 = 2 \times 3 \times 19$

$114 \div 1 = 114$

,

$114 \div 114 = 1$

$114 \div 2 = 57$

$114 \div 3 = 38$

$114 \div 19 = 6$

....

114 is a double prime

Note

Prime × prime = double prime

Prime × double prime = double prime

Double prime × double prime = double prime

4 Index laws

Definition

If $a^n = m$. a and n are positive integers

A is called the base and n is called the index

If $b > c$ Then $a^b > a^c$

Example 4-1 $2^9 = 512$

2 is called the base and 9 is called the index

Example 4-2 $9^2 = 81$

9 is called the base and 2 is called the index

Example 4-3 $a^n = a \times a \times a \times \ldots$ to n factors   •

$$2^5 = 2 \times 2 \times 2 \times 2 \times 2$$

...to 5 factors

Smallest exponent and greatest exponent

If $b > c$ Then $a^b > a^c$

a^b is the greatest exponent and a^c is the smallest Exponent

Example 4-4 $2 < 2^2 < 2^7$

2 is the smallest exponent

2^7 is the greatest exponent

Note $2^1 = 2$

5 Highest Common Factor.(H.C.F.)
Lowest Common Multiple.(L.C.M.)

Remember

H.C.F.

= Smallest exponent of common prime multiply

L.C.M.

= Greatest exponent of all prime multiply

Example 5-1

Find the H.C.F. of 12, 24, 60 and 90

Solution

$12 = 2^2 \times 3$

$24 = 2^3 \times 3$

$60 = 2^2 \times 3 \times 5$

$90 = 2 \times 3^2 \times 5$

Common prime   2,   3

Smallest exponent   2,   3

Multiply $2 \times 3 = 6$

So   H.C.F. = 6

(Multiplier is the factor of product)

Note   a.   The factor of each number ?

b.   The common factor of all numbers ?

c.   the H.C.F. of all numbers ?

Understand ?

Example   5- 2

Find the L.C.M. of 12, 24, 60 and 90

Solution

$12 = 2^2 \times 3$

$24 = 2^3 \times 3$

$60 = 2^2 \times 3 \times 5$

$90 = 2 \times 3^2 \times 5$

All prime   2,   3,   5

Greatest exponent 2^3, 3^2, 5

Multiply $2^3 \times 3^2 \times 5 = 120$

So   L.C.M. = 360

Note   a.   The multiple of each number ?

            b,   The common multiple of all number ?

            c.   The L.C.M. of all number ?

Understand ?

6   Application

Example 6-1   Reduce $\dfrac{456789}{456987}$

Solution

$$\frac{456789}{456987} = \frac{3\times43\times3541}{3\times23\times37\times179} = \frac{43\times3541}{23\times37\times179} = \frac{152263}{152329}$$

Example 6-2   Compare $\dfrac{123}{456}$ and $\dfrac{345}{678}$

Solution find the L.C.M. of 456 and 678

$$456 = 2^3 \times 3 \times 19$$

$$678 = 2 \times 3 \times 113$$

C.L.M. = $2^3 \times 3 \times 19 \times 113$

$$\frac{123\times113}{456\times113} = \frac{13899}{L.C.M.}$$
$$\frac{345\times2^2\times19}{678\times2^2\times19} = \frac{26220}{L.C.M.}$$

$$13899 \ < \ 26220$$

So $\dfrac{123}{456} < \dfrac{345}{678}$

Example 6-3   Find

a.   Common factor (C.F.)

b.   L.C.F. and

c.   H.C.F. of 462, 546 and 714

Solution

Minimum $= 462 = 2 \times 3 \times 7 \times 11$

$546 = 2 \times 3 \times 7 \times 13$

Maximum $= 714 = 2 \times 3 \times 7 \times 17$

C. F.

↓

2,   3,   7, 2×3, 2×7, 3×7, $2 \times 3 \times 7$

$\downarrow$

2, 3, 7, 6, 14, 21, 42

So

a. C.F. 2, 3, 7, 6, 14, 21, 42

b. L.C.F. = 2

c . H.C.F. = 42

(Smallest exponent of common prime multiply)

Note

L.C.F.	H.C. F.	Min		Max
$\downarrow$	$\downarrow$	$\downarrow$		$\downarrow$
..2..........24......		462...	546...	714...

Understand ?

Example 6-4   Find

a. Common factor (C.F.)

b. L.C.F. and

c. H.C.F. of 3 and 15

Solution

Minimum $= 3$

Maximum $= 15 = 3 \times 5$

C.F.$= 3$

So

a. C.F. = 3

b. L.C.F. = 3

c. H.C.F. =3

(Smallest exponent of common prime multiply)

Note

Min	Max
C.F	$\downarrow$

L.C.F.

H.C.F.

$\downarrow$

3...15.......

L.C.F. = C.F. = H.C.F. = 3 = Min < Max

Understand ?

Example 6-5   A cuboid of length 90 cm, breadth 60 cm and height 24 cm. To divide it evenly into cube, each as large as possible. What will be the length of the side of each cube ?

Analysis

The largest possible length is just the H.C.F. of 90 ,60 and 24

Solution

$24 = 2^3 \times 3$

$60 = 2^2 \times 3 \times 5$

$90 = 2 \times 3^2 \times 5$

common prime 2, 3

Smallest exponent   2,   3

Multiply: $2 \times 3 = 6$

$H.C.F. = 6$

So   The side of each cube is 6 cm

Example 6-6   Each area of square A and square B are $4m^2$ and $6m^2$ respectively. Find the smallest number of each square so that whole areas of each square are the same.

Solution

$4 = 2^2$

$6 = 2 \times 3$

All prime   2,   3

Greatest exponent 2^2, 3

Multiply $2^2 \times 3 = 12.(m^2)$

So

Smallest number of square A

$$= \frac{L.C.M.}{Area\ of\ square\ A} = \frac{12}{4} = 3$$

Smallest number of square B

$$= \frac{L.C.M.}{Srea\ of\ square\ B} = \frac{12}{6} = 2$$

Note   Common multiple

12,   24,   36,   48,   60,   72

$12 = 4 \times 3 = 6 \times 2$

$24 = 4 \times 6 = 6 \times 4$

$36 = 4 \times 9 = 6 \times 6$

$48 = 4 \times 12 = 6 \times 8$

$60 = 4 \times 15 = 6 \times 10$

$72 = 4 \times 18 = 6 \times 12$

… … … … … … … … … ..

Do you understand ?

1 正整數

定義

下述數

1, 2, 3, 4, 5, 6, 7, 8, 9, 10

11, 12, 13, 14, 15, 16, 17, 18, 19, 20

....

101,102,103,104,105,106,107,108,109,110

111,112,113,114,115,116,117,118,119,120

....

名為 正整數

下述各數

1,2,3,5,7,11,13,17,19,.....101,103,107,109,113

同是 一質數

下述各數

4,6,8,9,10,12,14,15,16,18,20,.....118,119........

同是一 (雙)重質數

所以 正整數包括 質數 和 重質數

注意

以上考慮 僅基于下述內容之原因

2 因數

定義

乘數 是 積的因數

例 2-1

$120 = 8 \times 15 = 2^3 \times 3 \times 5$

8, 15, 2, 3, 5　同是 120 的因數

有時 一個數包含有 多個因數 所以 求出所有因數并非容易
事實上

$$120 = 2 \times 2^2 \times 3 \times 5 = 2 \times 4 \times 3 \times 5$$

$$C_4^1 + C_4^2 + C_4^3 + C_4^4 = 4 + 6 + 4 + 1 = 15$$

$$C_4^1 = 4 \rightarrow 2, 4, 3, 5$$

$$C_4^2 = 6 \rightarrow 2 \times 4, \quad 2 \times 3, \quad 2 \times 5$$

$$4 \times 3, \quad 4 \times 5, \quad 3 \times 5$$

$\rightarrow \quad 8, \quad 6, \quad 10, \quad 12, \quad 20, \quad 15$

$C_4^3 = 4 \rightarrow 2 \times 4 \times 3, 4 \times 3 \times 5, 3 \times 5 \times 2, 5 \times 2 \times 4$

$\rightarrow \quad 24, \quad 60, \quad 30, \quad 40$

$C_4^4 = 1 \rightarrow 120$

所以 2,4,3,5,8,6,10,12,20,15,24,60,30,40,120　　同是 120 的因數

記住 一個數含有多個因數 求出每一個因數并非容易

3 質數

定義

僅只有被 1 和自身 兩個正整數 整除的正整數 名為質

能被 1 自身 和其它正整數 整除的正整數 名為 (雙)重質數

例 3-1　$5 = 1 \times 5$　$5 \div 1 = 5$　$5 \div 5 = 1$

5 是一個質數

例 3-2-　$113 = 1 \times 113$　　　$113 \div 1 = 1$

　　$113 \div 113 = 1$

113 是一個質數

例 3-3　$10 = 2 \times 5$　$10 \div 1 = 10$　$10 \div 10 = 1$

　　$10 \div 2 = 5$　$10 \div 5 = 2$

10 是一個重質數

例 3-4　　$114 = 2 \times 3 \times 19$

$114 \div 1 = 114$　　$114 \div 114 = 1$

$114 \div 2 = 57$　　$114 \div 3 = 38$

$114 \div 19 = 6$

114 是一個重質數

注意

　質數 乘 質數 = 重 質數

　質數 乘 重質數 = 重質數

　重質數 乘 重質數 = 重質數

4   指數定律

定義

如果 $a^n = m.$ a 和 n 同是一正整數

則 a 名為 底數   n 名為 指數

又   如果 b > c   則 $a^b > a^c$

例 1-1 $2^9 = 512$

2 名為 底數 9 名為 指數

例 2- 2 $9^2 = 81$

9 名為 底數 2 名為 指數

例 3- 3. $3a^n = a \times a \times a \times \dots n 個 a\ 相乘$

$2^5 = 2 \times 2 \times 2 \times 2 \times 2$

5 個 2 相乘

冪最小者   冪最大者

如果   b > c $a^b > a^c$

則 a^b 名為冪最大者

a^c 名為冪最小者

例 3- 4. $2 < 2^2 < 2^7$

2   名為冪最小者

2^7 名為冪最大.

注意 $2^1 = 2$

5　　最大公約數　最小公倍數

記住

最大公約數 ＝ 公有質數　冪最小者　相乘

最小公倍數 ＝ 所有質數　冪最大者　相乘

例 5- 1

求　12, 24, 60 和 90 的　最大公約數

解　$12 = 2^2 \times 3$

$24 = 2^3 \times 3$

$60 = 2^2 \times 3 \times 5$

$90 = 2 \times 3^2 \times 5$

公有質數　2,　3

冪最小者 2,　3

相乘　$2 \times 3 = 6$

所以　最大公約數 ＝ 6

(乘數是積的因數)

注意

a.　　每一個數的約數 ？

b.　　所有數的公約數 ？

c.　所有數的最大公約數 ？

明白嗎 ？

例 5- 2

求 12, 24, 60 和 90 的最小公倍數

解

$12 = 2^2 \times 3$

$24 = 2^3 \times 3$

$60 = 2^2 \times 3 \times 5$

$$90 = 2 \times 3^2 \times 5$$

所有質數　2,　3,　5

冪最大者　2^3,　3^2,　5

相乘　$2^3 \times 3^2 \times 5 = 360$

所以　最小公倍數 = 360

注意

a.　每一個數的倍數 ？

b.　所有數的公倍數 ？

c.　所有數的最小公倍數 ？

明白嗎 ？

6　　運用

例 6-1.　化簡 $\frac{456789}{456987}$

解

$$\frac{456789}{456987} = \frac{3 \times 43 \times 3541}{3 \times 23 \times 37 \times 179} = \frac{43 \times 3541}{23 \times 37 \times 179} = \frac{152263}{152329}$$

例 6-2　比較大小　$\frac{123}{456}$ 和 $\frac{345}{678}$　求 456 和 678　　的最小公倍數

$456 = 2^3 \times 3 \times 19$

$678 = 2 \times 3 \times 113$

最小公倍數 $= 2^3 \times 3 \times 19 \times 113$

$$\frac{123 \times 113}{456 \times 113} = \frac{13899}{\textit{最小公倍數}}$$

$$\frac{345 \times 2^2 \times 19}{678 \times 2^2 \times 19} = \frac{26220}{\textit{最小公倍數}}$$

$13899 < 26220$

所以　$\frac{123}{456} < \frac{345}{678}$

例 6-3　　求 462, 546 和 714 的

a.　　公約數

b.　　最小公約數

c.　　最大公約數

解

最小數　　$462 = 2 \times 3 \times 7 \times 11$

　　　　　　$546 = 2 \times 3 \times 7 \times 13$

　　　　　　$714 = 2 \times 3 \times 7 \times 17$

　　　　　公約數

　　　　　　↓.

2,　3,　7,　2 × 3, 2 × 7, 3 × 7, 2 × 3 × 7

↓

2,　3,　7,　6,　　14,　21,　　42

所以

a.　公約數　2, 3, 7, 6, 14, 21, 42

b.　最小公約數是 2

c.　最大公約數是 42

(公有質數　冪最小者相乘)

注意

最小約數　最大公約數　　最小數　最大數

　　2　　　　　24　　　　462　　　714

明白嗎 ？

例 6-4　　求 3 和 15 的

a.　公約數

b.　 最小公約數

c.　 最大公約數

解

最小數 = 3

最大數 = 15 = 3 × 5

公約數 = 3

所以

　a.　公約數 = 3

　b.　最小公約數 = 3

　c.　最大公約數 = 3

(公有質數　冪最小者相乘)

注意

最小數　　　　　　　最大數

最小公約數 ＝ 公約數 ＝ 最大公約數

＝ 最小數 ＜ 最大數

　　　　　　↓　　　　　　　　↓

.....................3..15...........

明白嗎 ？

例 6-5　　一長方體長 90cm 闊 60cm 高 24cm 均分為盡可能大的正方體 求每一
正方體的邊長 ？

最大可能的邊長正是 90,60,24 的最大公約數

解

$24 = 2^3 \times 3$

$60 = 2^2 \times 3 \times 5$

$90 = 2 \times 3^2 \times 5$

公有質數　2,　3

冪最小者　2,　　3

相乘　$2 \times 3 = 6$

最大公約數 ＝ 6

所以正方體的邊長是 6cm

例 6-6　　正方形 A, B 的面積分別是 $4m^2, 6m^2$. 求正方形 A 和 B 的總面積相等時
正方形 A 和 B 的最小個數

解

$4 = 2^2$

$6 = 2 \times 3$

所有質數 2,　3

冪最大者　2^2, 3

相乘: $2^2 \times 3 = 12(m^2)$

所以

正方形 A 的最小個數

$$= \frac{最小公倍數}{正方形A\ 的面積} = \frac{12}{4} = 3$$

正方形 B 的最小公個數

$$= \frac{最小公倍數}{正方形B的面積} = \frac{12}{6} = 2$$

注意　　公倍數

12,　24,　36,　48,　60,　72..........

$12 = 4 \times 3 = 6 \times 2$

$24 = 4 \times 6 = 6 \times 4$

$36 = 4 \times 9 = 6 \times 6$

$48 = 4 \times 12 = 6 \times 8$

$60 = 4 \times 15 = 6 \times 10$

$72 = 4 \times 18 = 6 \times 12$

....

明白嗎 ?